W0187684

This book belongs to:

....

....

....

HODDER CHILDREN'S BOOKS
First published in Great Britain in 2024
by Hodder & Stoughton

1 3 5 7 9 10 8 6 4 2

Text copyright © Adam Frost, 2024
Illustrations copyright © Mike Byrne, 2024

Adam Frost and Mike Byrne have asserted their right under
the Copyright, Designs and Patents Act 1988, to be identified
as the author and illustrator respectively of this work.
All rights reserved. A CIP catalogue record for this book
is available from the British Library.

PB ISBN 978-1-526-36568-2
E-book ISBN 978-1-444-97514-7

Printed in China

MIX
Paper | Supporting
responsible forestry
FSC® C104740
FSC
www.fsc.org

Hodder Children's Books
An imprint of Hachette Children's Group
Part of Hodder & Stoughton Limited
Carmelite House
50 Victoria Embankment
London, EC4Y 0DZ

An Hachette UK Company
www.hachette.co.uk
www.hachettechildrens.co.uk

ADAM FROST   MIKE BYRNE

SHARKS
LOVE
SCIENCE

Hodder
Children's
Books

DO YOU LOVE SCIENCE?

Being a scientist is one of the best jobs ever. You get to **EXPLORE** the world around you, ask all the **QUESTIONS** and make mind-blowing **DISCOVERIES!**

And one of the best places to explore is the sea, the most scientifically spectacular environment on Earth.

So, are you ready to SWIM with SHARKS?
And paddle with PENGUINS? And DIVE with
DOLPHINS? If so, this is the book for YOU!

SWIMMING WITH SHARKS

Picture a shark – what do you see? A big, grey oblong with sharp teeth? Well, think again! Sharks come in all SHAPES and SIZES!

A **hammerhead shark** looks like it's escaped from a toolbox! Its eyes are on each end of its wide head.

A **bull shark** has the classic shark shape. They're called bull sharks because they like to headbutt other fish!

The **bonnethead shark** is the only shark that eats vegetables. It likes to nibble on seagrass.

Living in the depths of the ocean, a **frilled shark** could be mistaken for a slithery sea serpent.

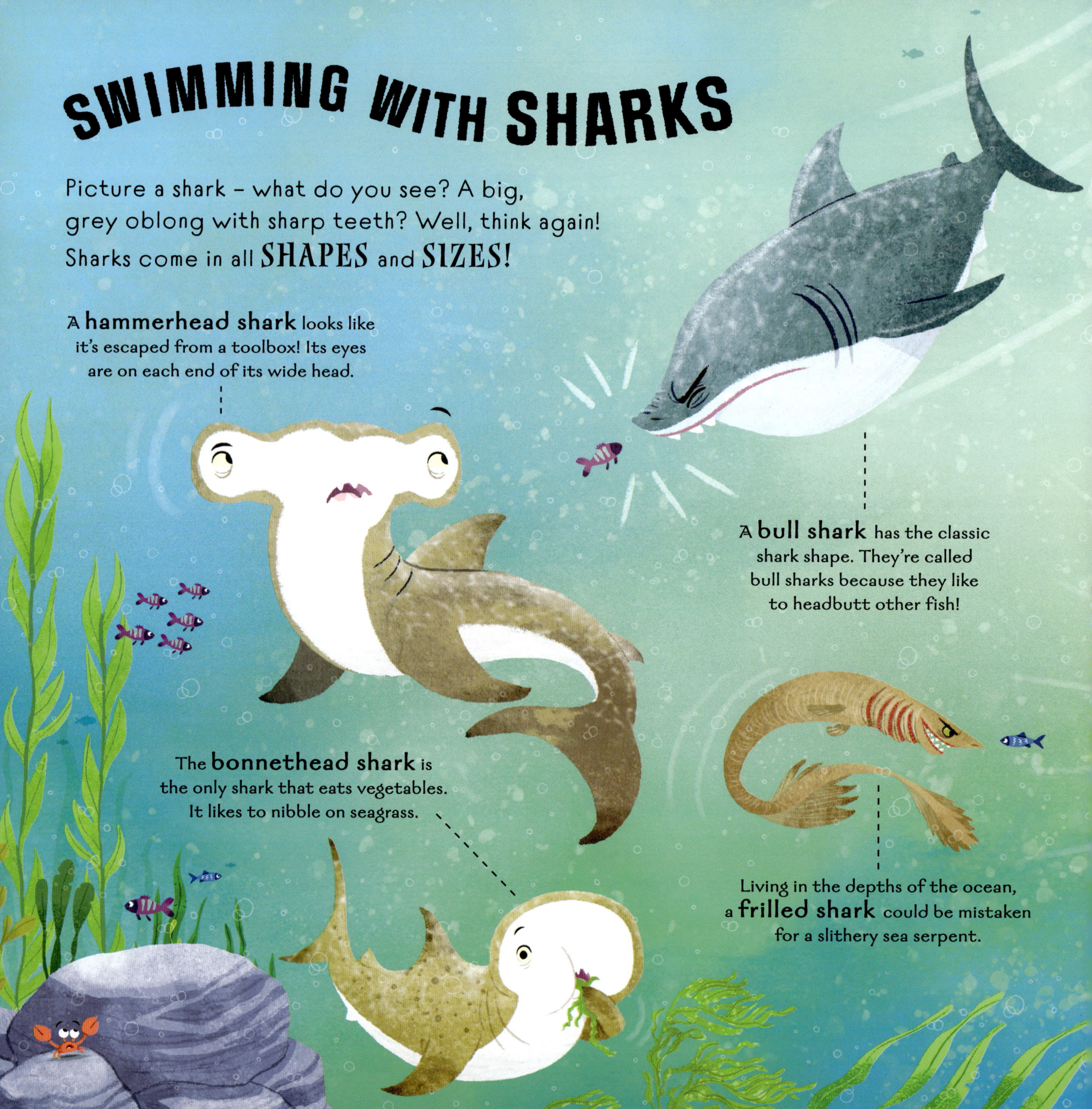

1. Which kind of shark glows in the dark?

2. One type of shark on this page isn't a carnivore (a meat-eater). Which one?

A **saw-shark** looks like it has a chainsaw for a nose. It uses its spiky snout to slice up its dinner.

ANSWERS:
1. Lantern shark;
2. Bonnethead shark

A **thresher shark** has an extremely long tail that it uses like a whip!

Shark or giant vacuum cleaner? The **basking shark** swims around with its enormous mouth open, sucking down tiny creatures called plankton.

Ever heard of a glow-in-the-dark shark? Well, that's what **lantern sharks** do. Their blue-green glow attracts prey.

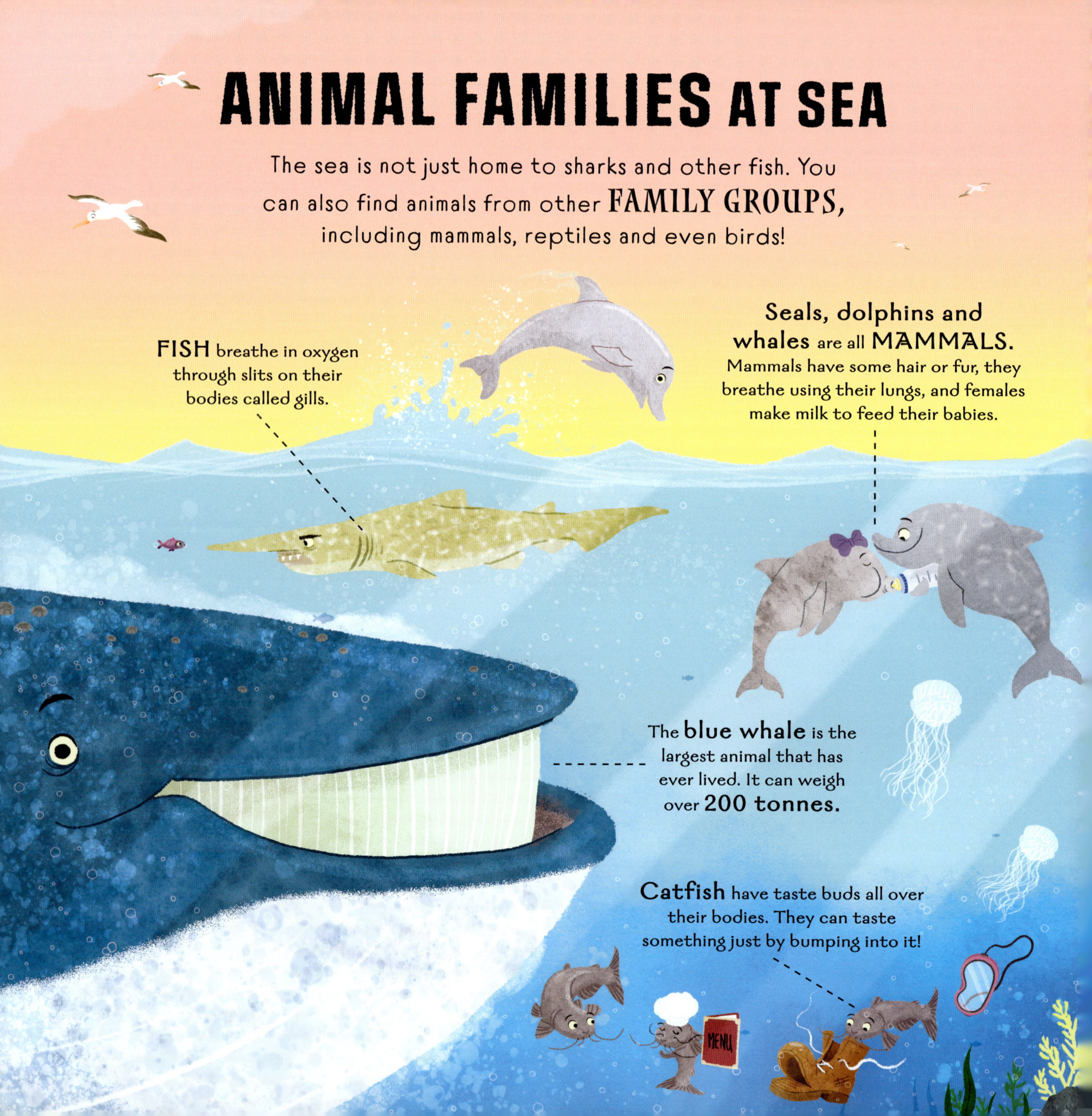

ANIMAL FAMILIES AT SEA

The sea is not just home to sharks and other fish. You can also find animals from other **FAMILY GROUPS**, including mammals, reptiles and even birds!

FISH breathe in oxygen through slits on their bodies called gills.

Seals, dolphins and whales are all **MAMMALS.** Mammals have some hair or fur, they breathe using their lungs, and females make milk to feed their babies.

The **blue whale** is the largest animal that has ever lived. It can weigh over **200 tonnes.**

Catfish have taste buds all over their bodies. They can taste something just by bumping into it!

MENU

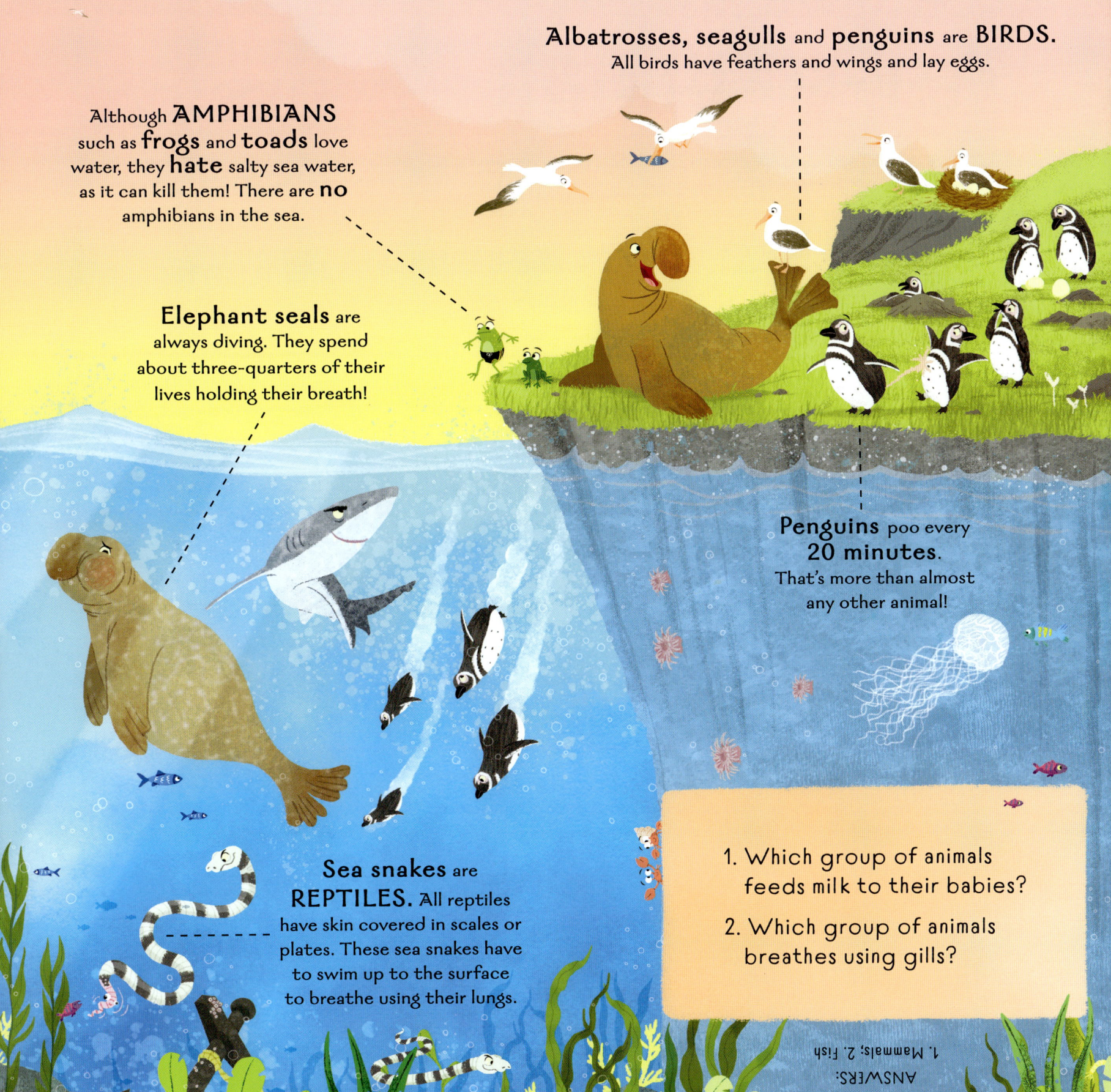

Albatrosses, seagulls and **penguins** are **BIRDS.**
All birds have feathers and wings and lay eggs.

Although **AMPHIBIANS** such as **frogs** and **toads** love water, they **hate** salty sea water, as it can kill them! There are **no** amphibians in the sea.

Elephant seals are always diving. They spend about three-quarters of their lives holding their breath!

Penguins poo every **20 minutes.** That's more than almost any other animal!

Sea snakes are **REPTILES.** All reptiles have skin covered in scales or plates. These sea snakes have to swim up to the surface to breathe using their lungs.

1. Which group of animals feeds milk to their babies?

2. Which group of animals breathes using gills?

ANSWERS: 1. Mammals; 2. Fish

DEEP-SEA DINING

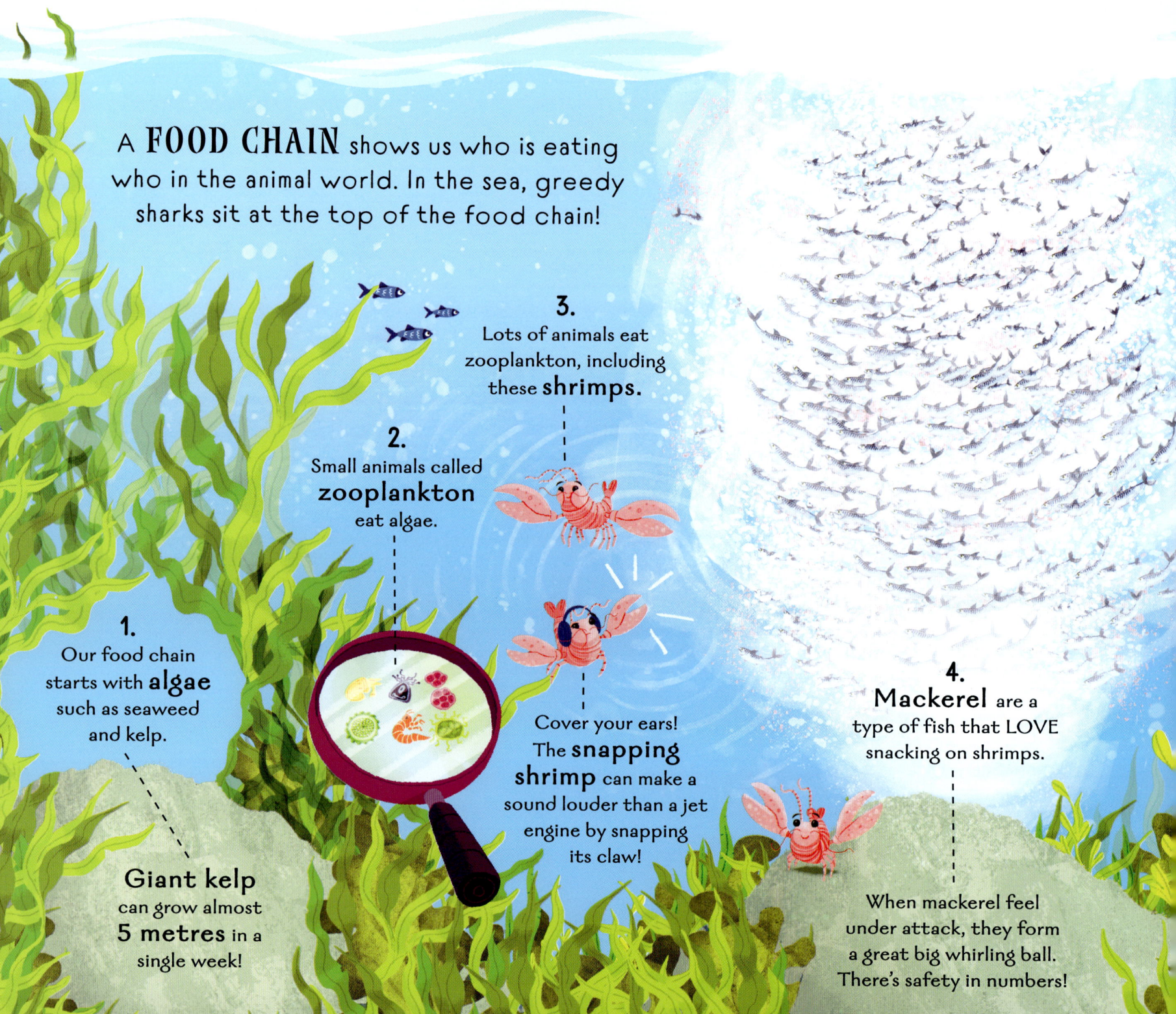

A **FOOD CHAIN** shows us who is eating who in the animal world. In the sea, greedy sharks sit at the top of the food chain!

3.
Lots of animals eat zooplankton, including these **shrimps.**

2.
Small animals called **zooplankton** eat algae.

1.
Our food chain starts with **algae** such as seaweed and kelp.

Giant kelp can grow almost **5 metres** in a single week!

Cover your ears! The **snapping shrimp** can make a sound louder than a jet engine by snapping its claw!

4.
Mackerel are a type of fish that LOVE snacking on shrimps.

When mackerel feel under attack, they form a great big whirling ball. There's safety in numbers!

A great white has **300** terrifying teeth in up to **seven** rows.

Large sharks only have one predator — **US!** Around **100 million** sharks are killed by humans every year.

6.
Great white sharks will feast on any fish, including tuna.

5.
Larger fish, such as **tuna,** eat mackerel.

Tuna vary hugely in size — from **80-centimetre**-long **skipjacks** (that's dog-sized!) to **4-metre**-long **Atlantic bluefins** (bigger than a horse!).

1. What's on the bottom of this food chain?

2. Which kind of fish is eating the mackerel?

ANSWERS:
1. The algae (seaweed and kelp);
2. Tuna

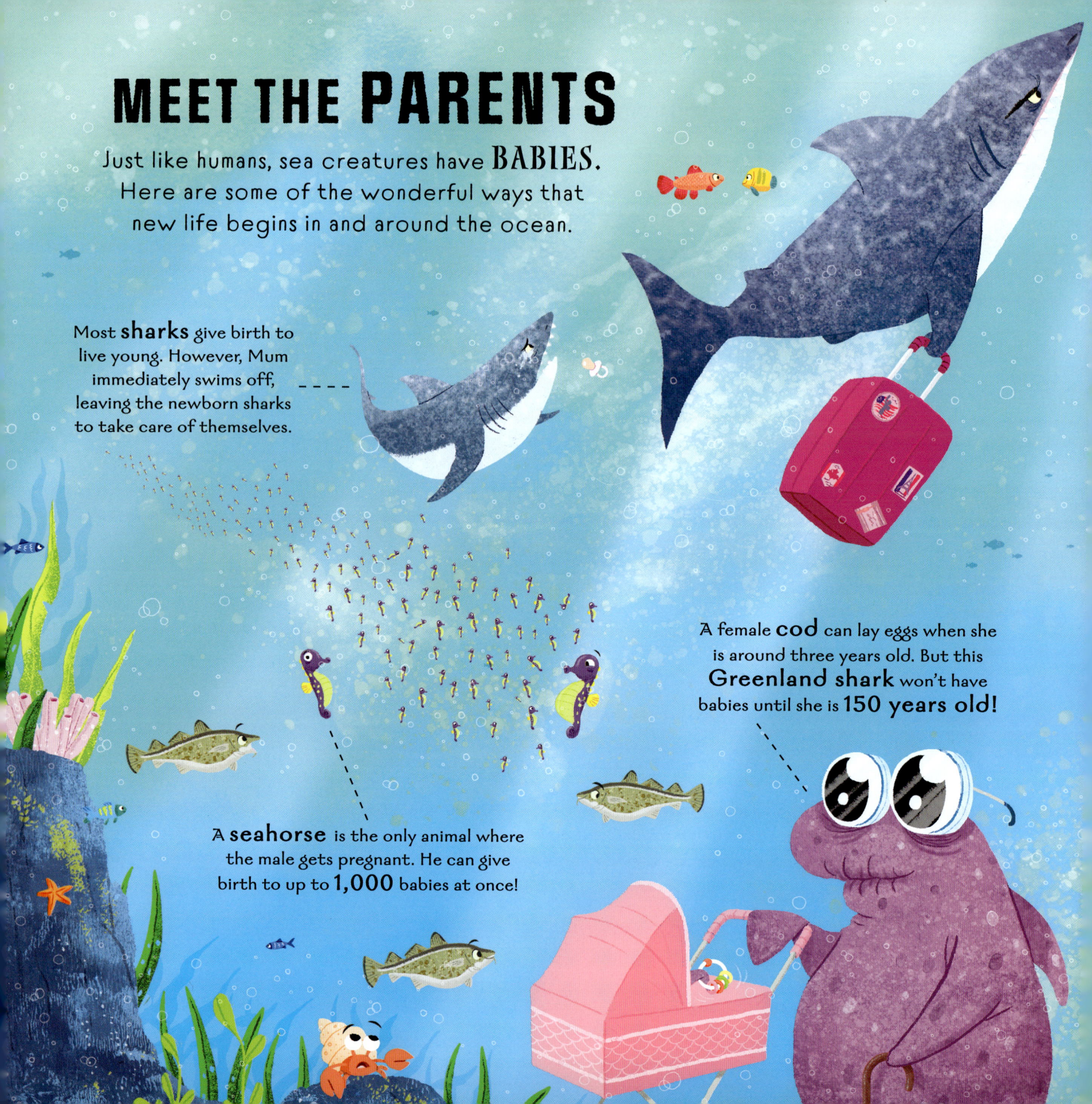

MEET THE PARENTS

Just like humans, sea creatures have BABIES. Here are some of the wonderful ways that new life begins in and around the ocean.

Most **sharks** give birth to live young. However, Mum immediately swims off, leaving the newborn sharks to take care of themselves.

A female **cod** can lay eggs when she is around three years old. But this **Greenland shark** won't have babies until she is **150 years old!**

A **seahorse** is the only animal where the male gets pregnant. He can give birth to up to **1,000** babies at once!

For two whole months, proud **emperor penguin** dads huddle together with eggs between their legs, protecting them against the cold.

When a female **dragon fish** lays eggs, the male keeps them safe in his mouth until they hatch. He cannot open his mouth to eat, in case the eggs fall out!

To attract a female, the male **pufferfish** creates beautiful patterns on the ocean floor. If a female likes his artwork, she'll lay her eggs in the middle of it.

Which fish don't lay eggs?

ANSWER: Sharks

UNDERWATER FORESTS

Did you know that there are forests of giant brown algae (kelp) under the sea? And just like RAINFORESTS on land, they are home to lots of plants and animals.

Kelp can be up to - - - 60 metres tall. That's over 10 times taller than the tallest giraffe!

- - - - Kelp forests need sunlight to grow, so they are only found in clear, shallow waters.

Animals that live in kelp forests include sea otters, giant octopuses and horn sharks.

Kelp forests are everywhere! They cover almost a quarter of the world's coastlines.

Humans love kelp too! It is grown in kelp farms and used in things such as yoghurt, toothpaste, ice cream and shampoo.

Sea otters often sleep holding hands to stop themselves drifting apart.

Horn sharks hide amongst the kelp during the day and venture out at night to munch on mussels and sea urchins.

1. Kelp can grow in the deepest oceans, even where there is no sunlight. True or false?

2. Can you name two everyday things you might use that contain kelp?

ANSWERS: 1. False; 2. Two of: yoghurt, toothpaste, ice cream and shampoo.

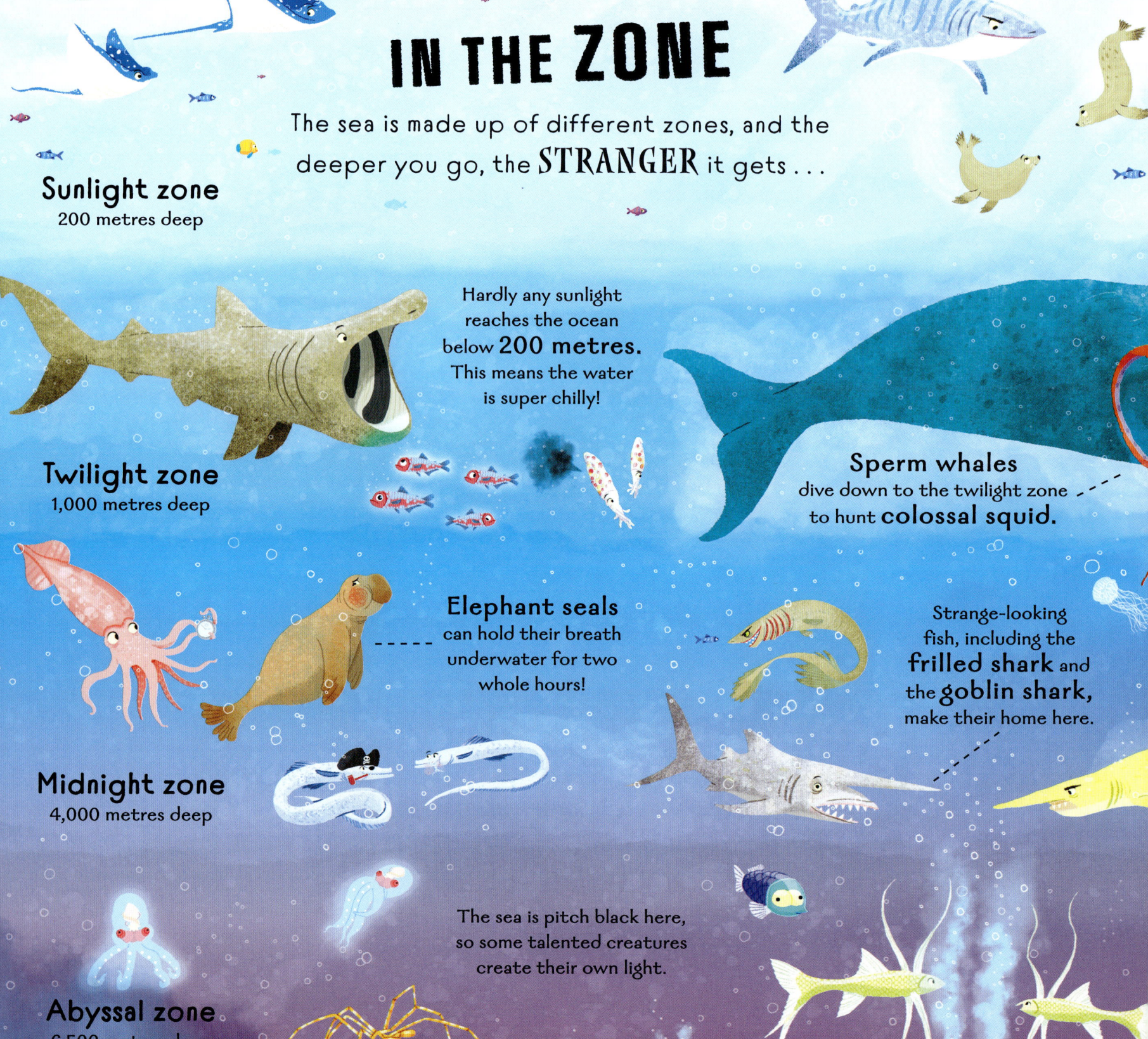

IN THE ZONE

The sea is made up of different zones, and the deeper you go, the STRANGER it gets...

Sunlight zone
200 metres deep

Hardly any sunlight reaches the ocean below **200 metres**. This means the water is super chilly!

Twilight zone
1,000 metres deep

Sperm whales dive down to the twilight zone to hunt **colossal squid**.

Elephant seals can hold their breath underwater for two whole hours!

Strange-looking fish, including the **frilled shark** and the **goblin shark**, make their home here.

Midnight zone
4,000 metres deep

The sea is pitch black here, so some talented creatures create their own light.

Abyssal zone
6,500 metres deep

About **90%** of all sea creatures can be found in the sunlight zone.

1. Which ocean zone is closest to the surface?
2. In which ocean zone would you find sperm whales hunting colossal squid?

ANSWERS:
1. Sunlight zone;
2. Twilight zone

The deepest-ever scuba dive (332 metres) was carried out by Ahmed Gabrin in 2014.

The wreck of famous passenger ship the *Titanic* lies **4,000 metres** deep in the midnight zone.

The **angler fish** that lurks here has a glowing light floating above its mouth to attract its prey.

At **10,994 metres** deep, the Mariana Trench is the deepest part of the ocean. Only three people have ever visited it — they had to use special submarines to get there!

LET'S STICK TOGETHER

Some sea animals like to live alone,
while others find strength in NUMBERS.

Dolphins sometimes form 'superpods' of around **1,000 dolphins** or 'megapods' of up to **100,000!**

Most sharks prefer to be alone. They can hunt well by themselves, and don't need any help!

A few kinds of shark are more sociable. **Reef sharks** hunt in packs.

Humpback whales live in family groups called pods. They sing songs that can be heard by other pods up to **8,000 kilometres** away.

Penguins are very sociable birds. The largest colony ever found was a group of **two million** chinstrap penguins.

Walruses live in herds. The males hang out in one herd and the females stick together in another.

When a shoal of **sardines** is under attack, they group together in a bait ball. These balls of spinning fish can be a whopping **20 metres** wide!

Herring live in large groups called schools. Schools of herring can contain up to **FOUR BILLION** fish.

1. What is a group of herring called?

2. How far can the song of a humpback whale travel through the water?

ANSWERS: 1. A school; 2. Up to 8,000 kilometres

NOW YOU SEE ME . . .

This shark is ready for its lunch but can't seem to find any food. Can you see through each creature's CAMOUFLAGE?

In forests of sea rods and pipe sponges, long, thin **trumpetfish** blend in perfectly.

The **leafy sea dragon** looks like a patch of seaweed. They have no predators — probably because they are so difficult to see!

The **MIMIC OCTOPUS** can change its shape to look like almost anything, including a lionfish. Lionfish are venomous and sharks won't eat them.

Octopus *in disguise*

Actual **lionfish**

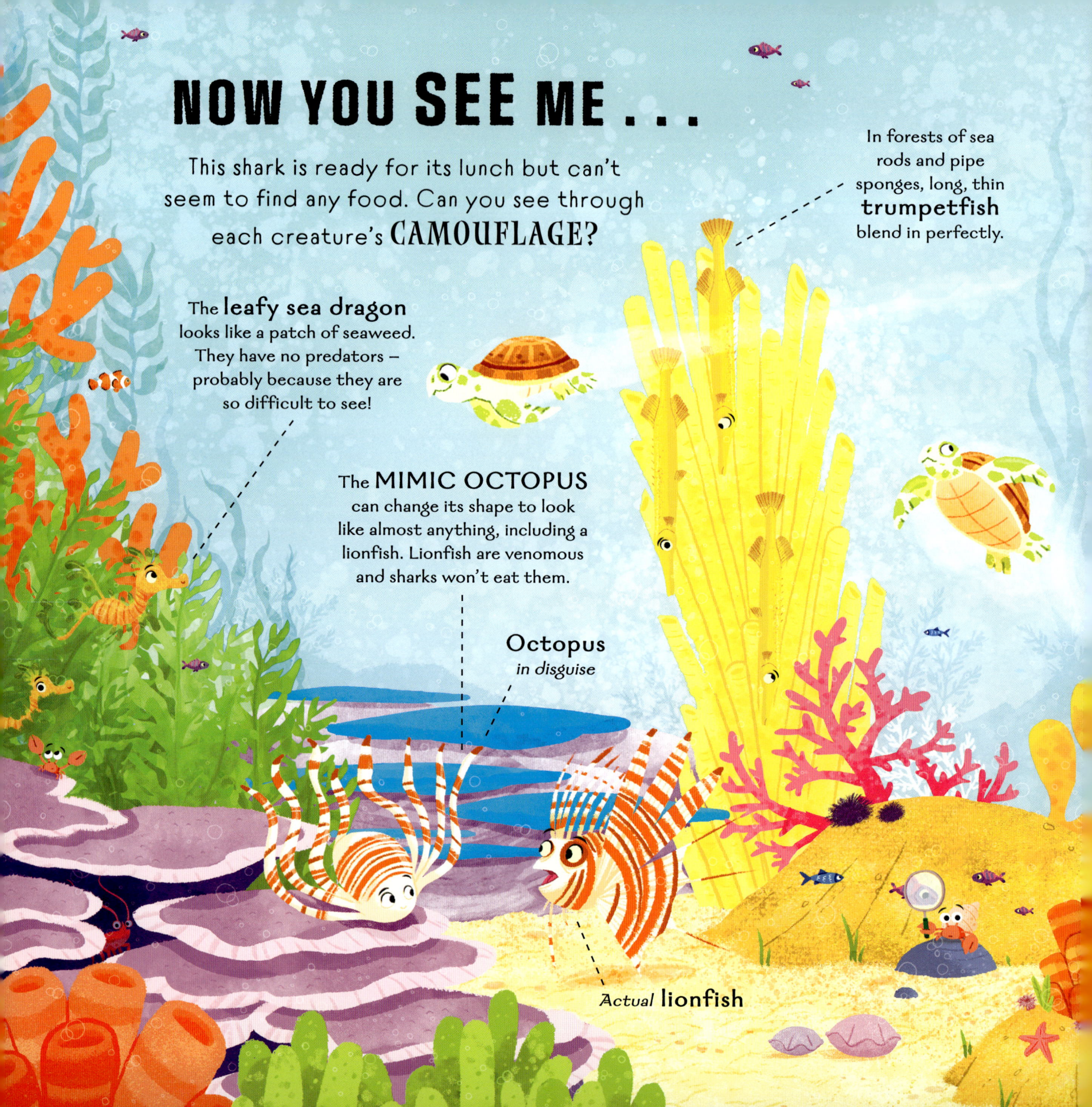

A **cuttlefish** is a master of disguise. In a split second, it can change the colour and texture of its skin.

The patterns on a **wobbegong shark's** back make it almost impossible to spot on a coral reef.

Could one of these rocks be a **stonefish?** Take care not to tread on them! Stonefish shoot out spines full of deadly poison.

Clever **decorator crabs** camouflage themselves with shells, sponges and seaweed.

1. Which kind of octopus can make itself look like a fish?

2. Which kind of crab covers itself with things it finds in the ocean?

ANSWER: 1. Mimic octopus; 2. Decorator crab.

OCEAN EXTREMES

Conditions at sea can get pretty scary. Waves, waterspouts, whirlpools and waterfalls can all cause CHAOS!

When **lightning** hits the sea, it spreads out across the water, electrocuting anything on or near the surface.

The largest wave ever recorded was in Alaska, USA. It was over **500 metres** tall. Yikes!

The largest **iceberg** ever seen was almost **100 kilometres** wide! It was spotted in the Southern Ocean.

The **wave** was over **four times** taller than the Statue of Liberty in New York City, USA!

About **90%** of an **iceberg** sits underwater.

Waterspouts stretch from the clouds down to the sea. They can move at speeds of up to **50 kilometres per hour.**

1. The world's biggest waterfall can be found under the sea. True or false?
2. How much of an iceberg sits underwater?

ANSWER: 1. True; 2. 90%

A **whirlpool** is a small area of the sea where the water moves round and round very quickly. Whirlpools can pull in anything that happens to be close by!

The biggest **waterfall** in the world is under the sea! Located just off the coast of Iceland, the water here hits an underwater ridge and then surges downwards for over **3 kilometres.**

TOUGH STUFF

It's not just plants and animals that are found in the sea. Lots of **ARTIFICIAL MATERIALS** drift around in the salty water too.

Wetsuits worn by divers are made of a tough rubber called **neoprene**.

Most fishing nets are made of a STRONG and STRETCHY plastic called **nylon**.

The Aquarius Reef Base in Florida, USA is the only underwater science lab in the world. It is made of **steel**, a metal that is incredibly strong.

Because **glass** does not rust like metal or dissolve in water, it can last for hundreds of years on the ocean floor.

Pumice is the only rock that floats. This is because it has pockets of air trapped inside it.

Unfortunately, humans throw things in the sea that don't belong there. The **Great Pacific Garbage Patch** contains around **1.8 trillion** pieces of trash.

Wood usually floats, but wooden shipwrecks sink because of all the metal things that are part of the ship, such as cannons, portholes and nails.

Sadly, tiger sharks sometimes eat this trash. The strangest things eaten by a tiger shark include a video camera and a suit of armour.

The ocean floor is covered in **rocks** and **stones** because they sink in water.

1. Which tough material is used for making wetsuits?

2. What is the only kind of rock that floats?

ANSWER:
1. Neoprene; 2. Pumice

THE BIG SCIENCE CHALLENGE

Are you ready to put your SCIENCE SKILLS to the test?

EXPERIMENT 1 – MAKE IT FLOAT!

It's so much easier to float in the sea than in a swimming pool. In this experiment, we'll find out why.

You will need:

o   Two fresh eggs
o   Two glasses
o   Salt
o   A tablespoon

1. Take a fresh egg and place it in a tall glass of water. Watch as the egg sinks to the bottom.

2. Take a second glass of water and add four tablespoons of salt. Stir the salt until it dissolves (disappears).

3. Now place another fresh egg in your salty water. What do you think will happen?

4. Record what you see. What do you think this shows?

Now you can see why salty sea water helps you to float. Salt makes the water denser than the egg – which means there is more 'stuff' in the water than in the egg.

BE SURE TO ASK A GROWN-UP TO HELP WITH THESE EXPERIMENTS!

EXPERIMENT 2 – MASTER OF MOVEMENT

Ever wondered how something as floppy as an octopus can move through the water so quickly? In this experiment, you'll become a master of jet propulsion!

You will need:

o A water balloon
o A washing-up liquid bottle top
o A bath or sink partly filled with water

1. Put the water balloon around the end of a tap and fill it up halfway with water. Make sure you have the washing-up liquid bottle top close by.

2. Fit the washing-up liquid bottle top inside the open end of the water balloon, with the open/close part poking out. Make sure the washing-up top is closed (pushed down).

3. Fill a sink or bath with at least 5 centimetres of water. Place your balloon 'octopus' in the water. Now open the washing-up liquid bottle top and watch what happens to your octopus.

As the water rushes out of the balloon, your 'octopus' darts through the water. This is what real-life octopuses do! They suck in water through holes in their bodies, and then push the water out through a funnel under their heads. They propel (move forwards) themselves using jets of water. This is jet propulsion!